NAHB-OSHA Jobsite Safety Handbook

Second Edition

National Association of Home Builders
U.S. Occupational Safety and Health Administration

NAHB-OSHA
Jobsite Safety Handbook, Second Edition

ISBN 0-86718-454-X
© 1999 by Home Builder Press®
of the National Association of Home Builders
of the United States of America

All rights reserved. No part of this book may be reproduced or utilized in any form or by any means, electronic or mechanical, including photocopying and recording, or by any information storage and retrieval system without permission in writing from the publisher.

Printed in the United States of America.

Library of Congress Cataloging-in-Publication Data

NAHB-OSHA jobsite safety handbook / National Association of Home Builders, U.S. Occupational Safety and Health Administration. 2nd ed.
   p. cm.
   ISBN 0-86718-454-X

   1. Building—United States—Safety measures. 2. House construction—United States—Safety measures. 3. Construction industry—Safety regulations-=-United States. I. National Association of Home Builders (U.S.) II. United States. Occupational Safety and Health Administration. III. Title: Jobsite safety handbook
TH443.N25 1998
690'.22—dc21
98-42711
CIP

Disclaimer

The information contained in this publication is not considered a substitute for any provisions of the Occupational Safety and Health Act of 1970 or for any standards written by OSHA.

This publication is designed to provide accurate and authoritative information in regard to the subject matter covered. It is sold with the understanding that the publisher is not engaged in rendering legal, accounting, or other professional service. If legal advice or other expert assistance is required, the services of a competent professional person should be sought.

                   —From a Declaration of Principles jointly adopted
                      by a Committee of the American Bar Association and a
                      Committee of Publishers and Associations

Quantity Discounts

Quantity discounts for individual Home Builder Press titles are available. Multi-title packages are also available for certain books. For further information, please contact—

   Director of Marketing
   Home Builder Press®
   National Association of Home Builders
   1201 15th Street, NW
   Washington, DC 20005-2800
   Check us out online at http://www.builderbooks.com
   or call (800) 368-5242, ext. 394; fax (202) 822-0391

1/99 Harlowe/P.A. Hutchinson 17,100

Acknowledgments

This second edition of the NAHB-OSHA *Jobsite Safety Handbook* resulted from a cooperative effort between the National Association of Home Builders and the U.S. Occupational Safety and Health Administration.

Numerous individuals and companies were integral to the development of this document. NAHB and OSHA wish to thank the following for their generous contribution of time and professional expertise in helping to develop this handbook:

Building Companies—Accent Decorators, Inc.; Aluminators, Inc.; Burja Construction; Cardinal Roofing; K&J Contractors; Snedden Brothers; and a special thanks to Winchester Homes for help with the photographs in this book.

Committee Members—Barry Larson, Chairman; Mike McMichael, Vice Chairman; James Anderson; David Asbridge; Bob Behlman; Pat Bridges; Steve Caporaso; Anthony Clatterbuck; Larry Franklin; Diane Glenn; Tony Goulet; Danny Graham; Jim Kuhn; Bob Masterson; Stuart Price; Leon Rogers; Craig Steele; Mike Thibodeaux; Bruce Thompson; Wesly Galyon; Bob Hanbury; and Chip Hughes.

Book Development—This handbook was developed and written under the direction of—

Kent W. Colton, Executive Vice President
 and Chief Executive Officer
National Association of Home Builders
1201 15th Street, NW
Washington, DC 20005-2800

Charles N. Jeffress, Assistant Secretary
Occupational Safety and Health
 Administration
United States Department of Labor
200 Constitution Avenue, NW
Washington, DC 20210

The *NAHB-OSHA Jobsite Safety Handbook* is a joint effort by the National Association of Home Builders and the Occupational Safety and Health Administration. The handbook is the second edition of the initial cooperative effort between NAHB and OSHA to assist builders and trade contractors in the residential construction industry.

Regina C. B. Solomon, CSP, then NAHB Director of Labor, Safety, and Health Services, prepared the first edition under the general direction of NAHB's Kent Colton, and OSHA's Joseph A. Dear (Assistant Secretary 1992–96). Solomon is now President of Aurus Safety Management, Inc., in Anderson, SC.

This handbook is designed to identify safe work practices and related OSHA requirements that have an impact on the most hazardous activities in the construction industry. Many detailed and lengthy requirements—such as the lead and asbestos standards—applicable to portions of the industry are not included in this handbook.

This handbook also does not replace any requirements detailed in the actual OSHA regulations for construction (Title 29 Code of Federal Regulations, Part 1926); the handbook should only be used as a companion to the actual regulations.

The main goal of the handbook is to explain in an easily understood language what builders can do to comply with safe work practices and some of the OSHA requirements. The goal of the handbook is to help the residential construction industry comply with OSHA standards while focusing on the most common hazards found on their jobsites.

If any inconsistency ever exists between the handbook and the OSHA regulations, the OSHA regulations (29 CFR 1926) will always prevail. This document should never be considered a substitute for any provisions of a regulation.

If you have any questions regarding the content of this handbook, please contact—

David D. DeLorenzo
Director, Labor, Safety, and Health Services
National Association of Home Builders
1201 15th Street, NW
Washington, DC 20005-2800
E-mail: ddelorenzo@nahb.com
(800) 368-5242, ext. 226

Contents

Introduction 1
Safety and Health Program Guidelines 2
Employee Duties 3
Employer Duties 4
Orientation and Training 4
Personal Protective Equipment 5
    Head Protection 6
    Eye and Face Protection 6
    Foot Protection 7
    Hand Protection 7
    Fall Protection 7
Housekeeping and Access at Site 8
Stairways and Ladders 9
Scaffolds and Other Work Platforms 13
    General 13
    Planking 14
    Scaffold Guardrails 16
Fall Protection 18
    Floor and Wall Openings 18
    Alternatives 18
    Work on Roofs 21
Excavations and Trenching 23
    General 23
    Foundations 26
Tools and Equipment 27
Vehicles and Mobile Equipment 29
Electrical 30
Fire Prevention 32

Figures

1. Worker with personal protective equipment   5
2. Clean jobsite   8
3. Properly guarded stairs   9
4. Two ways to secure the base of a ladder   10
5. Ladder for access to upper level   11
6. Proper angle for ladder and three-point contact   12
7a. and 7b. Scaffold footing, mudsill, and baseplate   14
8. Checklist for safe scaffold use   15
9. Safe fabricated frame scaffold   17
10. Properly erected pump jack scaffold   17
11. Guardrail for window opening   19
12. Guardrail around a floor opening   20
13. Correct height for guardrails and midrails   20
14. Safe work practices for truss work   21
15a. and 15b. Slide guards for fall protection   22
16. Slide guards for 7:12 pitch roof   23
17. Profile of a residential excavation   24
18. Trench box   25
19. Benched trench along a house foundation   26
20. Properly guarded power saw   28
21. Earth-moving equipment with safety devices   29
22. Extension cord protected by ground fault circuit interrupter   31
23. The PASS method   32
24. Safety can for flammable liquids   33

Introduction

The residential construction industry represents a significant percentage of the construction work force. Safe work practices of small building companies play an important part in reducing injuries and fatalities in the residential construction industry.

OSHA defined *residential construction* in the December 1995 "Interim Fall Protection Guidelines for Residential Construction" as "structures where the working environment, and the construction materials, methods, and procedures employed are essentially the same as those used for typical house (single-family dwelling) and townhouse construction. Discrete parts of a large commercial structure may come within the scope of this definition (for example, a shingled entranceway to a mall), but such coverage does not mean that the entire structure thereby comes within the terms of this definition."

This *Jobsite Safety Handbook* highlights the minimum safe work practices and regulations related to the major hazards and causes of fatalities occurring in the residential construction industry. The information presented in this handbook does not exempt the employer from compliance with all the requirements contained in Title 29 Code of Federal Regulations, Part 1926, any state or local safety laws and regulations and applicable standards for the residential construction industry. You should use the *Jobsite Safety Handbook* only as a general guide to safety practices.

For additional specific legal requirements and safety practices relevant to your particular job, you should rely on the specific regulations and generally accepted safe work practices that are accepted in your area.

Safety and Health Program Guidelines

Employers need to institute and maintain a company program of policies, procedures, and practices to protect their employees from, and help them to recognize, job-related safety and health hazards.

The company safety program should include procedures for the identification, evaluation, and prevention or control of workplace hazards, specific job hazards, and potential hazards that may arise.

An effective company safety program will include the following four main elements:

1. Management Commitment
The most successful company safety program includes a clear statement of policy by the owner, management support of safety policies and procedures, and employee involvement in the structure and operation of the program.

2. Worksite Analysis
An effective company safety program sets forth procedures to analyze the jobsite and identify existing hazards and conditions and operations in which changes might occur to create new hazards.

SAFETY AND HEALTH PROGRAM GUIDELINES

3. Hazard Prevention and Control
An effective safety program establishes procedures to correct or control present or potential hazards on the jobsite.

4. Safety and Health Training
Training is an essential component of an effective company safety program. The complexity of training depends on the size and complexity of the worksite as well as the characteristics of the hazards and potential hazards at the site.

Employee Duties

- ☑ Follow all safety rules
- ☑ Wear and take care of personal protective equipment
- ☑ Make sure all safety features for tools and equipment are functioning properly
- ☑ Don't let your work put another worker in danger
- ☑ Replace damaged or dull hand tools immediately
- ☑ Avoid horseplay, practical jokes, or other activities that create a hazard
- ☑ Don't use drugs or alcohol on the job
- ☑ Report any unsafe work practice and any injury or accident to your supervisor

Employer Duties

- ☑ Keep the workplace free from hazards
- ☑ Inform employees of how to protect themselves against hazards that cannot be controlled
- ☑ Conduct regular jobsite safety inspections
- ☑ Have someone trained in first aid on site if you have no emergency response service nearby

Orientation and Training

Each worker must receive safety orientation and training on applicable OSHA standards, company safety requirements, and/or have enough experience to do his/her job safely. You should evaluate this training occasionally to ensure proper understanding and implementation of the company safety requirements and OSHA standards.

☑ Personal Protective Equipment

Workers must use personal protective equipment, but it is not a substitute for taking safety measures. Workers still need to avoid hazards (Figure 1).

Figure 1. This worker is preparing to cut lumber while wearing the proper personal protective equipment. He is wearing a hard hat and safety glasses, and the saw is guarded correctly. His employer has determined that he should use hearing protection.

HEAD PROTECTION

- ☑ Workers must wear hard hats when overhead, falling, or flying hazards exist or when danger of electrical shock is present.

- ☑ Inspect hard hats routinely for dents, cracks, or deterioration.

- ☑ If a hard hat has taken a heavy blow or electrical shock, you must replace it even when you detect no visible damage.

- ☑ Maintain hard hats in good condition; do not drill; clean with strong detergents or solvents; paint; or store them in extreme temperatures.

EYE AND FACE PROTECTION

- ☑ Workers must wear safety glasses or face shields for welding, cutting, nailing (including pneumatic), or when working with concrete and/or harmful chemicals.

- ☑ Eye and face protectors are designed for particular hazards so be sure to select the type to match the hazard.

- ☑ Replace poorly fitting or damaged safety glasses.

FOOT PROTECTION

- ☑ Residential construction workers must wear shoes or boots with slip-resistant and puncture-resistant soles (to prevent slipping and puncture wounds).

- ☑ Safety-toed shoes are recommended to prevent crushed toes when working with heavy rolling equipment or falling objects.

HAND PROTECTION

- ☑ High-quality gloves can prevent injury.
- ☑ Gloves should fit snugly.
- ☑ Glove gauntlets should be taped for working with fiberglass materials.
- ☑ Workers should always wear the right gloves for the job (for example, heavy-duty rubber for concrete work, welding gloves for welding).

FALL PROTECTION

- ☑ Use a safety harness system for fall protection.
- ☑ Use body belts only as positioning devices—not for fall protection.

Housekeeping and Access at Site

- ☑ Keep all walkways and stairways clear of trash/debris and other materials such as tools and supplies to prevent tripping.

- ☑ Keep boxes, scrap lumber, and other materials picked up. Put them put in a dumpster or trash/debris area to prevent fire and tripping hazards (Figure 2).

- ☑ Provide enough light for workers to see and to prevent accidents.

Figure 2. The builder keeps this jobsite clean by using an onsite trash collection bin.

Stairways and Ladders

- ☑ Install permanent or temporary guardrails on stairs before stairs are used for general access between levels to prevent someone from falling or stepping off edges (Figure 3).

- ☑ Do not store materials on stairways that are used for general access between levels.

- ☑ Keep hazardous projections such as protruding nails, large splinters, etc. out of the stairs, treads and handrails.

Figure 3. Worker is walking up properly guarded steps.

10 STAIRWAYS AND LADDERS

- ☑ Correct any slippery conditions on stairways before they are used.

- ☑ Keep manufactured and job-made ladders in good condition and free of defects.

- ☑ Inspect ladders before use for broken rungs or other defects so falls don't happen. Discard or repair defective ladders.

- ☑ Secure ladders near the top or at the bottom to prevent them from slipping and causing falls.

- ☑ When you can't tie the ladder off, be sure the ladder is on a stable and level surface so it cannot be knocked over or the bottom of it kicked out (Figure 4).

- ☑ Place ladders at the proper angle (1 foot out from the base for every 4 feet of vertical rise, Figure 5).

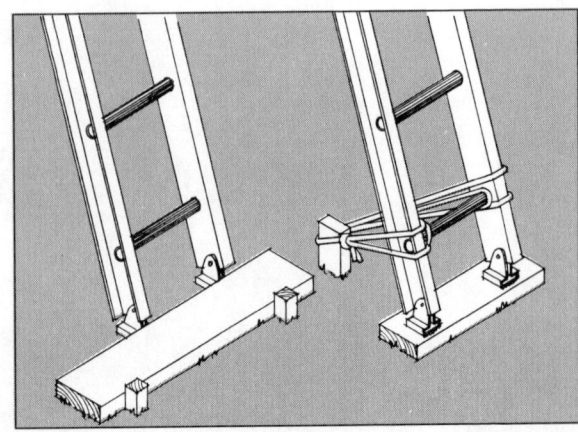

Figure 4. The drawing shows two ways to secure the base of a ladder to ensure proper footing.

STAIRWAYS AND LADDERS   11

- ☑ Extend ladders at least 3 feet above the landing to provide a handhold or for balance when getting on and off the ladder from other surfaces (Figure 5).

- ☑ Do not set up a ladder near passageways or high traffic areas where it could be knocked over.

- ☑ Use ladders only for what they were made and not as a platform, runway, or as scaffold planks.

- ☑ Always face the ladder and maintain 3 points of contact when climbing a ladder (Figure 6).

Figure 5. When ladders are used for access to an upper level they must extend at least 3 feet above the roof surface.

12 STAIRWAYS AND LADDERS

Figure 6. This worker is climbing a ladder set at the proper angle (4:1) with a three-point contact grip (two hands and one foot).

Scaffolds and Other Work Platforms

GENERAL

- ☑ Provide safe access to get on and off of scaffolds and work platforms safely. Use ladders safely (see Stairways and Ladders).

- ☑ Keep scaffolds and work platforms free of debris. Keep tools and materials as neat as possible on scaffolds and platforms. These practices will help prevent materials from falling and workers from tripping.

- ☑ Erect scaffolds on firm and level foundations (Figure 7a and 7b).

- ☑ Finished floors will normally support the load for a scaffold or work platform and provide a stable base.

- ☑ Place scaffold legs on firm footing and secure from movement or tipping, especially on dirt or similar surfaces (Figures 7a and 7b).

- ☑ Erect and dismantle scaffolds only under the supervision of a competent person.

- ☑ Each scaffold must be capable of supporting its own weight and 4 times the maximum intended load.

- ☑ The competent person must inspect scaffolds before each use.

14 SCAFFOLDS AND OTHER WORK PLATFORMS

Figures 7a and 7b. Stable footings and mud sills for this scaffold ensure the stability of the work platform. In this example (right) the siding contractor actually had the base plate manufactured to penetrate the ground while stabilizing the pump jack poles.

- ☑ Use manufactured base plates or mud sills made of hardwood or equivalent to level or stabilize the footings. Don't use blocks, bricks, or pieces of lumber.
- ☑ Also see the checklist in Figure 8.

PLANKING

- ☑ Fully plank a scaffold to provide a full work platform or use manufactured decking. The platform decking and/or scaffold planks must be scaffold grade and must not have any visible defects.
- ☑ Keep the front edge of the platform within 14 inches of the face of the work.

Figure 8. Safe Scaffold Use

- ☑ **DO NOT** use damaged parts that affect the strength of the scaffold.
- ☑ **DO NOT** allow employees to work on scaffolds when they are feeling weak, sick, or dizzy.
- ☑ **DO NOT** work from any part of the scaffold other than the platform.
- ☑ **DO NOT** alter the scaffold.
- ☑ **DO NOT** move a scaffold horizontally while workers are on it, unless it is a mobile scaffold and the proper procedures are followed.
- ☑ **DO NOT** allow employees to work on scaffolds covered with snow, ice, or other slippery materials.
- ☑ **DO NOT** erect, use, alter, or move scaffolds within 10 feet of overhead power lines.
- ☑ **DO NOT** use shore or lean-to scaffolds.
- ☑ **DO NOT** swing loads near or on scaffolds unless you use a tag line.
- ☑ **DO NOT** work on scaffolds in bad weather or high winds unless the competent person decides that doing so is safe.
- ☑ **DO NOT** use ladders, boxes, barrels, or other makeshift contraptions to raise your work height.
- ☑ **DO NOT** let extra material build up on the platforms.
- ☑ **DO NOT** put more weight on a scaffold than it is designed to hold.

- ☑ Extend planks or decking material at least 6 inches over the edge or cleat them to prevent movement. The work platform or planks must not extend more than 12 inches beyond the end supports to prevent tipping when workers are stepping or working.

- ☑ Be sure that manufactured scaffold planks are the proper size and that the end hooks are attached to the scaffold frame.

SCAFFOLD GUARDRAILS

- ☑ Guard scaffold platforms that are more than 10 feet above the ground or floor surface with a standard guardrail. If guardrails are not practical, use other fall protection devices such as safety harnesses and lanyards (Figure 9).

- ☑ Place the top rail approximately 42 inches above the work platform or planking with a midrail about half that high at 21 inches (Figure 10).

- ☑ Install toe boards if other workers will be below the scaffold.

SCAFFOLDS AND OTHER WORK PLATFORMS **17**

Figure 9. Workers stand on a fabricated frame scaffold. They have ladder access to the top of the scaffold (out of view); guardrails, cross bracing, and complete planking to prevent falls. The workers are also wearing hard hats and using eye protection.

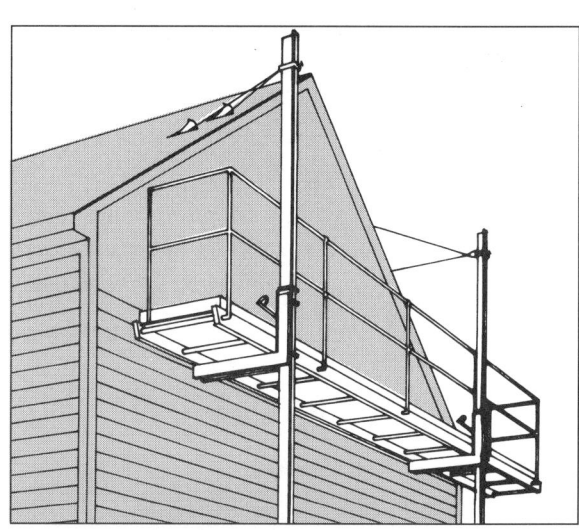

Figure 10. This pump jack scaffold was erected properly with guardrails and roof connectors. Because of the pump jack's limited strength, only two workers or up to 500 pounds are allowed on the unit.

Fall Protection

FLOOR AND WALL OPENINGS

- [✓] Install guardrails around openings in floors and across openings in walls when the fall distance is 6 feet or more. Be sure the top rails can withstand a 200-pound load (Figures 11 and 12).

- [✓] Construct guardrails with a top rail approximately 42 inches high with a midrail about half that high at 21 inches (Figure 13).

- [✓] Install toe boards when other workers are to be below the work area.

- [✓] Cover floor openings larger than 2x2 inches with material to safely support the working load.

ALTERNATIVES

- [✓] Use other fall protection systems such as slide guards, roof anchors, or alternative safe work practices when a guardrail system cannot be used.

- [✓] Wear proper slip-resistant shoes or footwear to lessen slipping hazards.

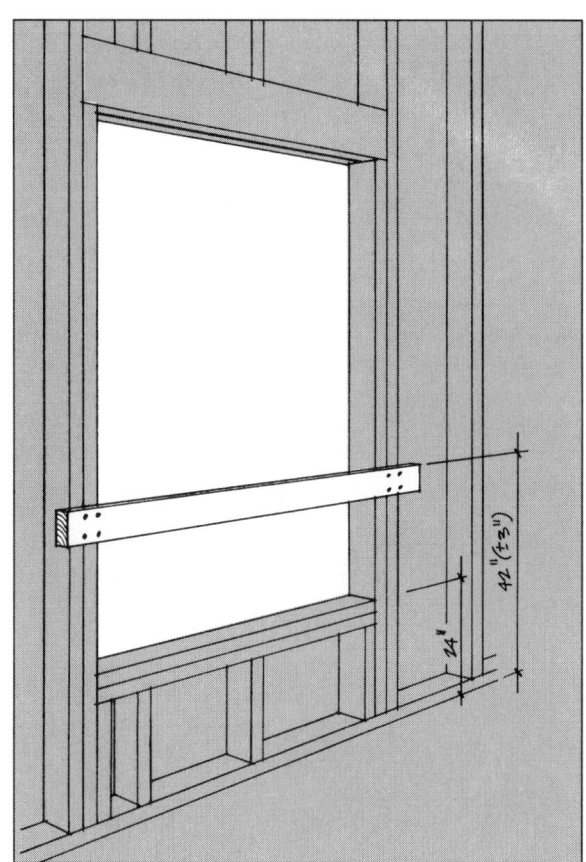

*Figure 11.
This window opening has a guardrail because the bottom sill height is less than 39 inches. Because the distance between the studs is less than 18 inches, no guardrails are needed between the studs.*

☑ Train workers in safe work practices before they perform work on foundation walls, roofs, trusses (Figure 14), or before they perform exterior wall erections and floor installations.

20 FALL PROTECTION

Figure 12. This photograph shows a proper guardrail around a floor opening.

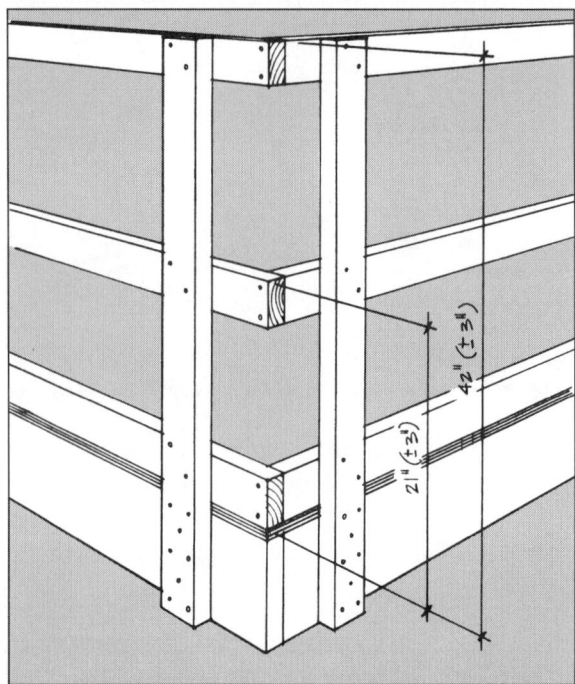

Figure 13. This drawing shows the correct height for guardrails and midrails—about 42 and 21 inches high respectively.

Figure 14. This worker uses a recognized safe work practice by standing on a work platform to secure the end of the roof truss.

WORK ON ROOFS

- ☑ Inspect for and remove frost and other slipping hazards before getting onto roof surfaces.

- ☑ Cover and secure all skylights and openings, or install guardrails to keep workers from falling through the openings.

- ☑ When the roof pitch is over 4:12 and up to 6:12, install slide guards along the roof eave after the first 3 rows of roofing material.

- ☑ When the pitch exceeds a 6:12 pitch, install slide guards along the roof eave after the first 3 rows of roofing material are installed and again every 8 feet up the roof (Figures 15a, 15b, and 16).

- ☑ Use a safety harness system with a solid anchor point on steep roofs with a pitch greater than 8:12 or if the ground-to-eave height exceeds 25 feet.

22 FALL PROTECTION

Figure 15a (top) and 15b. These photographs show properly installed slide guards along the roof eave. The slide guard is a roof bracket with a 2x6 at a 90-degree angle.

- ☑ Stop roofing operations when storms, high winds, or other adverse conditions create unsafe conditions.
- ☑ Remove or properly guard any impalement hazards.
- ☑ Wear shoes with slip-resistant soles.

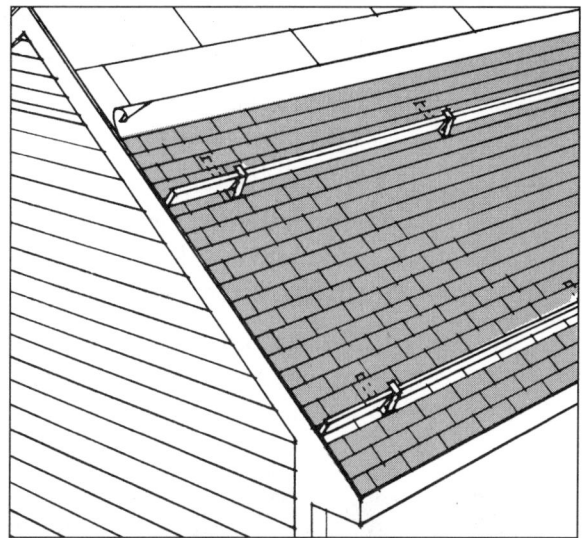

*Figure 16.
This 7:12 pitch roof
has properly installed
slide guards.*

Excavations and Trenching

GENERAL

- ☑ Find the location of all underground utilities by contacting the local utility locating service before digging.

- ☑ Keep workers away from digging equipment and never allow workers in an excavation when equipment is in use.

- ☑ Keep workers from getting between equipment in use and other obstacles and machinery that can cause crushing hazards.

- [✓] Keep equipment and the excavated dirt (spoils pile) back 2 feet from the edge of the excavation (Figure 17).

- [✓] Have a competent person conduct daily inspections and correct any hazards before workers enter a trench or excavation.

- [✓] Provide workers a way to get into and out of a trench or excavation such as ladders and ramps. They must be within 25 feet of the worker.

- [✓] For excavations and utility trenches over 5 feet deep, use shoring, shields (trench boxes), benching, or slope back the sides. Unless a soil analysis has been com-

Figure 17. The dotted line shows the profile of this excavation, as it was sloped at 1½ :1. Usually residential excavations are type C soil and will require such a slope. The spoils pile is at least 2 feet back from the edge of the excavation.

pleted, the earth's slope must be at least 1½ feet horizontal to 1 vertical (Figure 18).

- ☑ Keep water out of trenches with a pump or drainage system, and inspect the area for soil movement and potential cave-ins.

- ☑ Keep drivers in the cab and workers away from dump trucks when dirt and other debris are being loaded into them. Don't allow workers under any load and train them to stay clear of the backs of vehicles.

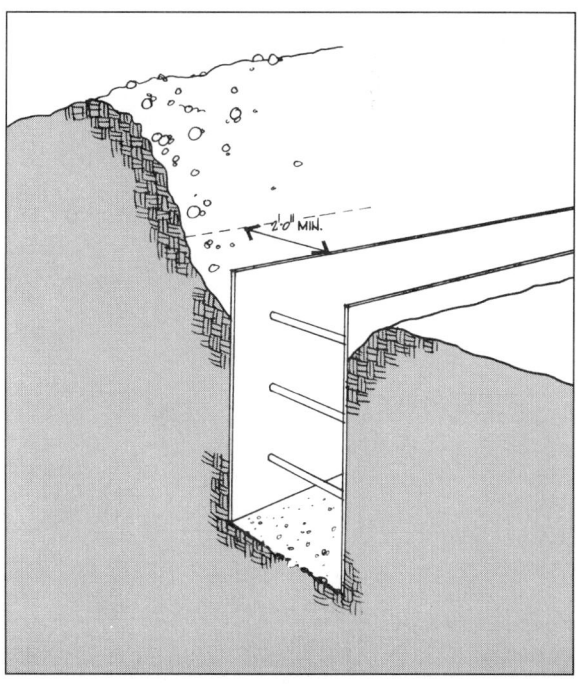

Figure 18. This trench box is being used correctly.

FOUNDATIONS

After the foundation walls are constructed, take special precautions to prevent injury from cave-ins in the area between the excavation wall and the foundation wall (Figure 19).

☑ The depth of the foundation/basement trench cannot exceed 7½ feet deep unless you provide other cave-in protection.

☑ Keep the horizontal width of the foundation trench at least 2 feet wide. Make sure no work activity vibrates the soil while workers are in the trench.

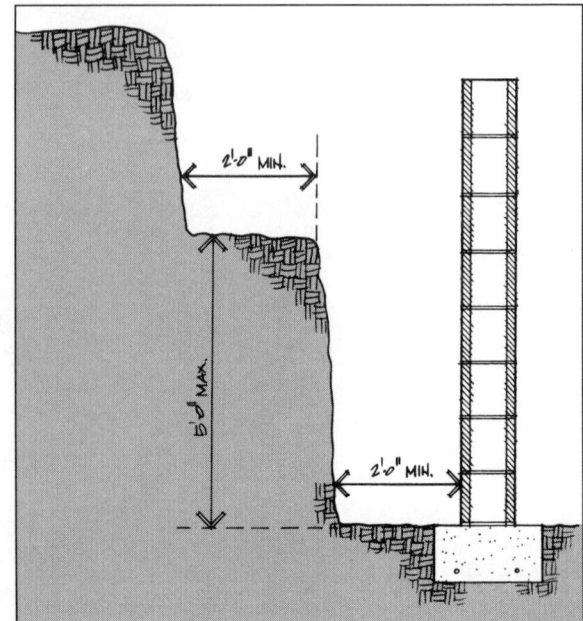

Figure 19. This drawing shows a properly benched trench along a house foundation.

- ☑ Plan the foundation trench work to minimize the number of workers in the trench and the length of time they spend there.
- ☑ Inspect the trench regularly for changes in the stability of the earth (water, cracks, vibrations, spoils pile). Stop work if any potential for cave-in develops and fix the problem before work starts again.

Tools and Equipment

- ☑ Maintain all hand tools and equipment in a safe condition and check them regularly for defects. Remove broken or damaged tools and equipment from the jobsite.
- ☑ Follow manufacturer's requirements for safe use of all tools.
- ☑ Use double insulated tools, or ensure that the tools are grounded.
- ☑ Equip all power saws (circular, skill, table, etc.) with blade guards.
- ☑ Make sure guards are in place before using power saws (Figure 20). Don't use power saws with the guard tied or wedged open.
- ☑ Turn off saws before leaving them unattended.
- ☑ Raise or lower tools by their handles, not by their cords.

Figure 20. This worker is using a power saw that has all moving parts, including the saw blade, properly guarded.

- ☑ Don't use wrenches when the jaws are sprung to the point of slippage. Replace them.
- ☑ Don't use impact tools with mushroomed heads. Replace them.
- ☑ Keep wooden handles free of splinters or cracks and be sure the handles stay tight in the tool.
- ☑ Workers using powder-activated tools must receive proper training prior to using the tools.
- ☑ Always be sure that hose connections are secure when using pneumatic tools.
- ☑ Never leave cartridges for pneumatic or powder-actuated tools unattended. Keep equipment in a safe place, according to the manufacturer's instructions.
- ☑ Require proper eye protection for workers.

Vehicles and Mobile Equipment

- ☑ Train workers to stay clear of backing and turning vehicles and equipment with rotating cabs.
- ☑ Be sure that all off-road equipment used on site is equipped with rollover protection (ROPS) (Figure 21).
- ☑ Maintain back-up alarms for equipment with limited rear view or use someone to help guide them back.
- ☑ Be sure that all vehicles have fully operational braking systems and brake lights.
- ☑ Use seat belts when transporting workers in motor and construction vehicles.
- ☑ Maintain at least a 10-foot clearance from overhead power lines when operating equipment.

Figure 21. This worker has been properly trained to operate this piece of equipment, and it is equipped with the appropriate safety devices.

- [x] Block up the raised bed when inspecting or repairing dump trucks.
- [x] Know the rated capacity of the crane and use accordingly.
- [x] Ensure the stability of the crane.
- [x] Use a tag line to control materials moved by a crane.
- [x] Verify experience or provide training to crane and heavy equipment operators.

Electrical

- [x] Prohibit work on new and existing energized (hot) electrical circuits until all power is shut off and a positive Lockout/Tagout System is in place.
- [x] Don't use frayed or worn electrical cords or cables.
- [x] Use only 3-wire type extension cords designed for hard or junior hard service. (Look for any of the following letters imprinted on the casing: S, ST, SO, STO, SJ, SJT, SJO, SJTO.)
- [x] Maintain all electrical tools and equipment in safe condition and check regularly for defects.
- [x] Remove broken or damaged tools and equipment from the jobsite.

ELECTRICAL

☑ Protect all temporary power (including extension cords) with ground fault circuit interrupters (GFCIs). Plug into a GFCI-protected temporary power pole, a GFCI-protected generator, or use a GFCI extension cord to protect against shocks (Figure 22).

☑ Don't bypass any protective system or device designed to protect employees from contact with electrical current.

☑ Locate and identify overhead electrical power lines. Make sure that ladders, scaffolds, equipment, or materials never come within 10 feet of electrical power lines.

Figure 22. The generator is a temporary power source so the builder has used a cord protected by a ground fault circuit interrupter (GFCI) to protect workers against electrocution. If the extension cord was plugged into an outlet in the house, it would still need a GFCI because the extension cord provides temporary power.

Fire Prevention

- ☑ Provide fire extinguishers near all welding, soldering, or other sources of ignition.
- ☑ Keep fire extinguishers easy to see and reach in case of an emergency.
- ☑ Provide one fire extinguisher within 100 feet of employees for each 3,000 square feet of building. (Figure 23).
- ☑ Don't store flammable or combustible materials in areas used for stairways or exits.
- ☑ Avoid spraying of paint, solvents, or other types of flammable materials in rooms with poor ventilation. Build-up of fumes and vapors can cause explosions or fires.

THE PASS METHOD

Pull the pin.

Aim the nozzle.

Squeeze the lever.

Sweep the nozzle.

Figure 23. Employees should be trained to use the PASS method to extinguish a fire.

FIRE PREVENTION **33**

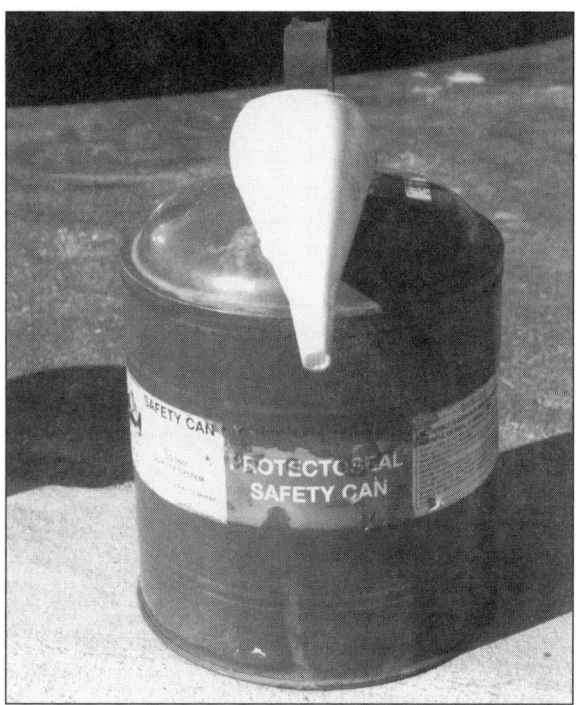

Figure 24. Gasoline and other flammable liquids need to be stored in a safety can.

- ☑ Store gasoline and other flammable liquids in a safety can outdoors or in an approved storage facility (Figure 24).
- ☑ Don't store LP gas tanks inside buildings.
- ☑ Keep temporary heaters at least 6 feet away from any LP gas container.
- ☑ Ensure that leaks or spills of flammable or combustible materials are cleaned up promptly.

Notes